The Plant Identikit

This booklet extends the Tasmanian series of plant identikits. The series was started by the Conservation Commission of the Northern Territory and has since been extended to a number of locations in Victoria, New South Wales and South Australia.

The booklets aim to encourage familiarity with the common plants of a region. They are inexpensive and pocket sized so that they may be used in the bush. In writing this booklet we hope that we are able to encourage a love for Tasmania's native plants.

In the same series

Wildflowers of Mt Wellington—Plant Identikit (1988,1997)
Alpine Wildflowers of Tasmania—Plant Identikit (1989, 1991, 1995)
Woodland Wildflowers of Tasmania—Plant Identikit (1990, 1993, 2006)
Rainforest Plants of Tasmania—Plant Identikit (1992, 1999)
Orchids of Tasmania—Plant Identikit (1993)
Coastal Plants of Tasmania—Plant Identikit (2003)

Acknowledgements

This booklet is a product of the combined resources of many members of Australian Plants Society Tasmania Inc. (formerly the Society for Growing Australian Plants Tasmanian Region Inc.) The artists are listed on the page opposite. Margaret Allan was responsible for coordination of the project and field trips. Heather Gulline combined with Margaret Allan to provide botanical advice and criticism of the text while it was being written. Gladys and Harry Dodson read the text carefully and advised on consistent style and correct usage of English. We received permission to use the name Plant Identikit from the Conservation Commission of the Northern Territory. The Department of Parks, Wildlife and Heritage gave us permission to collect plant material from various reserves for the artists' use. Dr Tony Orchard allowed access to the Tasmanian Herbarium to research distributions and flowering times.

Phil Collier, May 1990.

This second edition updates nomenclature and adds some more suggested locations for finding specimens. My thanks to Alex Buchanan of the Tasmanian Herbarium for his assistance with this. It has also been necessary to re-do some of the illustrations because of changes to printing techniques and the unavailability of some of the original artwork. My sincere thanks to Noel Kerrison for his patience with resetting the text and to Gladys Dodson who again did the proof reading.

Christine Howells, 2nd Edition, August 2006.

Introduction

Tasmania has many fine national parks and reserves in lowland areas, especially near the coast. These offer many recreational activities including the study of nature. We aim to increase your enjoyment of the wildflowers by providing an introduction to those that may be seen near the roadside and beside the walking tracks.

This booklet describes wildflowers that are found in lowland woodlands and heathlands, mostly in the north and east of the State. It does not include those plants which are most common on coastal rocks and dunes (see Coastal Plants of Tasmania Plant Identikit). Nor does it specialise in the woodlands found in highland areas such as the Central Plateau. Most of the wildflowers in the booklet are common, but we have included a few which are less common and occur naturally only in Tasmania, or are of special interest for other reasons.

Although the national parks and reserves are the strongholds of wildflowers which are easy to see, there are many remnants of natural vegetation throughout the north and east of the State. These may be found on road verges, in uncleared paddocks, and especially in state forests. It is not necessary to be a hardy walker to study the wildflowers in many of these places.

The booklet starts with a number of introductory sections. The first describes wooded and heathland habitats in more detail, and suggests a number of ideas for visiting these areas of the State. Next is a section on how to use this booklet, including details of how we have laid out the description of each species. At the end is an index of common names and of scientific names.

Our aim is to encourage the study of our native plants. Such a small booklet cannot expect to cover all the different kinds that can be found in woodlands and heathlands. The section *Further Reading* will give additional advice to those who become 'hooked' on the beauty of our native plants.

X 0.5

Leptorhynchos nitidulus

Woodlands and Heathlands in Tasmania

Woodland and Heathland Vegetation

Woodland and heathland vegetation types are widespread in Tasmania, extending well into the highlands. This book serves as an introduction to the woodland and heathland plants on dry sites in lowland Tasmania. This includes many parts of the Midlands and eastern parts of the State.

The terms woodlands and heathlands have a technical meaning to biogeographers. In Australia woodlands are usually defined as places dominated by woody plants over 2m tall with significant gaps in the canopy's foliage cover. Heathlands are places with shrubs less than 2m tall. On fertile sites, in the lowlands of Tasmania, it is the lack of sufficient rainfall which creates conditions suitable for woodlands to establish. On infertile sites and exposed coastal sites trees disappear completely leaving only low shrubs to form a heathland.

Fire is an important element in the existence of most lowland woodlands and heathlands. Many plants are highly adapted to surviving fire by one of a number of methods. Some shrubs and trees have dormant buds under their bark or in their woody rootstock. These buds may produce new branches after the fire has passed. Other shrubs are killed by fire but possess woody fruits which open after the fire has passed to release their seed. Other species have a very hard covering on their seeds and can survive underground for many years. Some herbaceous species, like the orchids, die down to an underground tuber during the dry fire-prone months of the summer.

Woodlands and heathlands are rarely uniform over large areas. Soil type and aspect have a significant influence on the plants that are found on a site. An extreme example of aspect can be seen in gullies where streams and rivers offer ample moisture which supports lush vegetation compared with the nearby slopes. A few plants in this booklet are specialised to the damper habitats in the drier parts of the State.

Visiting woodland and heathland areas in Tasmania

Since white settlement, the woodlands and heathlands of Tasmania have been extensively modified by farming and forestry. However, significant areas of native vegetation still exist. These may be found especially in national parks, reserves and some state forests.
The national parks which are best for viewing lowland wildflowers are probably Freycinet and Rocky Cape. Both offer nearby accommodation and extensive walking tracks. Freycinet National Park

is at its best in late winter and early spring. Rocky Cape National Park is wetter and the wildflowers may be seen through to the late spring. Narawntapu National Park, Mt William National Park and Waterhouse Protected Area are also worth a visit for their wildflower display. Several of the parks offer nature trails and explanatory leaflets.

Many of the large towns offer small reserves where wildflowers may be seen. Good examples are Henry Somerset Orchid Reserve near Devonport, Granite Point Coastal Reserve at Bridport, and at Launceston the Cataract Gorge and Kate Reed Reserve. Hobart can offer Knocklofty Reserve and Ridgeway Park as well as East Risdon Nature Reserve, Waverley Park Nature Reserve at Bellerive and Peter Murrell Reserve near Blackmans Bay. Different types of habitat in different reserves offer different wildflowers.

The Midlands is the heart of the grazing industry in Tasmania and has few remnants of its former vegetation. Those which do remain contain several species of wildflower which are rare elsewhere in the State. A notable area is Tunbridge tip (also known as Township Reserve) which is designated as a refuge. Epping Forest still has a significant area uncleared and part of this is the Tom Gibson Reserve.

The central part of the east coast around Swansea has a distinctive climate. There are long periods with little rain and infrequent bursts of heavy rain. The rocky dolerite hills in this area have developed a distinctive flora with a number of rare plants restricted to Tasmania. Cherry Tree Hill between Swansea and Bicheno, and the eastern end of the Lake Leake Road have interesting roadside vegetation.

Another region of the east coast with distinctive wildflowers is the Tasman Peninsula in the south-east. There are several scenic headlands such as Cape Hauy and Cape Raoul which support extensive heathlands. Both have walking tracks which offer a mixture of wildflowers and outstanding coastal scenery. Perhaps the most outstanding headland is Cape Pillar. The wildflowers there are at their best in November, but the walk to the Cape is better undertaken as an overnight camping expedition.

Although we recommend walking in the bush to observe wildflowers, nearly all of the plants illustrated in this booklet can be seen a few paces from a car door at parks, reserves or elsewhere in Tasmania. We wish the reader luck when searching for wildflowers.

How to use the Plant Identikit

The pictures in this booklet are designed to provide the primary method of identification of any plant in flower. As far as possible they have been ordered according to the colour of their flowers. The white flowers occur first, then pink and red. Following these are the creamy-yellow and yellow flowers. Towards the end, blue and violet flowers are grouped together, with green flowers and pines appearing last. Some species have been included mainly for their fruit colour. Such fruits are illustrated clearly, but for consistency will be found in the section of their flower colour.

If a search for the flower colour is unsuccessful, then it may be from a species not illustrated. It is still likely that a similar species will be found here. In this case it is worth searching for similar-looking flowers. For example there are several members of the heath family illustrated, many of which have tubular flowers. Some flowers are regular (all petals look the same), while others are irregular (asymmetric). If something similar is found, the notes headed *Related Species* may provide useful information.

Plants which are not in flower present difficult problems, not only to users of this booklet, but to botanists generally. There are still useful clues to be found about a plant's identity. Look at the leaves. Are they opposite each other, or alternate? Are they serrated or smooth on the edge? Are they hairy or shiny? These clues will eliminate many of the species illustrated here without the need to read a word of the text.

The text of the booklet is organised according to the explanation on p.8. The section *Distinguishing Features* should serve to confirm or reject an identification, so that is all that needs to be read during a preliminary scan of the pages.

Throughout the booklet we have placed emphasis on the scientific names, rather than common names. The challenge of becoming familiar with these names can form part of the wildflower experience. They are designed to be helpful since common names vary across different regions of the State and in different States. Also many plants have no common name, and we have resisted the temptation to invent them or publish names not in common use.

The scientific name is in two parts. The first part is the name of the genus (plural genera), and the second part is the species (plural species). Plants in the same genus are closely related. Additionally the family name is given. There are often a number of genera in a family, and these genera are also closely related. After a little practice, the genera and families become useful organising principles. If it is not possible to recognise the species of a plant, often the genus or family may be obvious. This will help in the search for a firm identification. Important families include the Myrtaceae (gum trees, tea-trees etc.), Epacridaceae (the heath family containing many small tough shrubs) and Asteraceae (the daisy flowers). To be completely precise we have occasionally included names more specific than the species name. These are subspecies (ssp.) or variety (var.).

One problem with scientific names is that they are apt to change as more study is carried out on Australian plants. This makes life difficult for the amateur and professional botanist alike. We have endeavoured to use the most up-to-date name throughout.

Wherever possible, we have avoided the use of botany's less well-known technical terms, of which there are a great many. The few that we felt compelled to use are followed by a brief explanation in parentheses. The reader is reminded that a flower consists of an outer ring of sepals, which are usually green and leaf-like, although sometimes coloured like the petals, especially in the orchids and lilies. Next is usually a ring of coloured petals, and inside these are the reproductive parts called the stamens (male) and carpel(s) (female). Sometimes what appears to be a flower is a collection of small flowers. This is true of daisy and wattle flowers. Although we have tried to be precise in our descriptions, we have also endeavoured not to mislead by being overly pedantic.

Further Reading

This booklet provides only an introduction to the woodland and heathland plants of Tasmania. There are several other reference books describing a larger selection of Tasmania's and Australia's native plants, many of which include woodland and heathland plants.

Picture Books

These books are designed as an introduction to some of the more interesting and attractive plants of an area. They are well illustrated with photographs, paintings or drawings.

M. Cameron (ed.), *Guide to Flowers and Plants of Tasmania,* Reed, Sydney, 3rd ed. 2000.

J. B. Kirkpatrick & S. Backhouse, *Native Trees of Tasmania,* Pandani Press, Sandy Bay, 2004.

Wildflowers of Mt Wellington—Plant Identikit, APST Inc., Hobart, 1997.
L. Gilfedder et al., *The Nature of the Midlands,* Midlands Bushweb, Longford, 2003.
J. Whiting et al., *Tasmania's Natural Flora,* Tasmania's Natural Flora Editorial Committee, Ulverstone, 2004.

Comprehensive Flora

For the serious botanist, descriptions of Tasmania's flowering plants may be found in the following books.

W. M. Curtis & D. I. Morris, *The Student's Flora of Tasmania,* Part 1, 2nd ed., Government Printer, Hobart, 1975.
W. M. Curtis, *The Student's Flora of Tasmania,* Parts 2, 3 and 4A, Government Printer, Hobart, 1963, 1967 and 1979.
W. M. Curtis & D. I. Morris, *The Student's Flora of Tasmania,* Part 4B, St David's Park Publishing, 1994.

Technical reports and papers

Some lowland areas of Tasmania have been studied intensively. Various reports and papers discuss woodland and heathland plant communities and many give check-lists of plants found in specific areas.

M. J. Brown & H. J. Bayly-Stark, *Vegetation of Maria Island,* National Parks and Wildlife Service Technical Report, Hobart,1979.
M. J. Brown & F. Duncan, *Dry Sclerophyll Vegetation in Tasmania,* National Parks and Wildlife Service Technical Report, 1985.
M.J. Brown & F. Duncan, *Vegetation of Tasman Peninsula,* Royal Society of Tasmania, 1989.
M. J. Firth, *Flora of Rocky Cape,* Records of the Queen Victoria Museum, No. 33, Launceston, 1969.
S. Harris & J. B. Kirkpatrick, *The Vegetation of Schouten Island, Tasmania,* Papers and Proceedings of the Royal Society of Tasmania, 1982.
J. B. Kirkpatrick, *The Disappearing Heath,* Tasmanian Conservation Trust, Hobart, 1977.
J. B. Kirkpatrick & S. Harris, *The Disappearing Heath Revisited,* Tasmanian Environment Centre Inc., Hobart,1999.
J. B. Kirkpatrick, M. J. Brown & A. Moscal, *Threatened Plants of the Tasmanian East Coast,* Tasmanian Conservation Trust, Hobart, 1988.
J. B. Kirkpatrick, L. Gilfedder & R. Fensham, *City Parks and Cemeteries,* Tasmanian Conservation Trust, Hobart, 1988.
J. B. Reid et al., *Vegetation of Tasmania,* Flora of Australia Supplementary Series No. 8, Australian Biological Resources Study, 1999.

Species name Common name
Family name

Distinguishing Features
A few characteristics which serve to distinguish this species from most of the others present in Tasmanian woodlands and heathlands.

Description
Further characteristics which serve to confirm the identification of this species. This should be read in conjunction with the distinguishing features for a fuller description.

Flowering Period
Gives the typical flowering period of the species in woodlands and heathlands.

Distribution
A description of the known woodland and heathland distribution for each species. Mostly the distributions cover eastern and northern areas of the State where rainfall is not excessive. Some species also occur in the heathlands on the west coast and some even extend into alpine areas.

Locality
This is designed for people wishing to locate plants of the species being described. When space allows we have included more than one locality. Many plants described as common will be found in many places including roadsides.

Cultivation Notes
Information about how the species may be propagated and cultivated in the garden. This has been gathered from the experience of APS members and notes from native plant publications. Some species are not commonly cultivated and the corresponding advice is vague. The comments in this section do not give or imply permission for plant materials to be collected from any area. Plant material should not be taken from many places especially national parks and other reserves.

Related Species
Further information which relates to the described species and its close relatives.

Wurmbea dioica ssp. **dioica** Early Nancy

Liliaceae

Distinguishing Features

A small lily with about three to five white flowers which have a purple band near the base of each petal.

Description

A small plant with an erect flower stem which has a few leaves surrounding its base. Most plants are either male or female. The male flowers have six obvious stamens, while female flowers have a purple ovary in the centre of the flower. The purple bands on the petals indicate the position of the organ which produces nectar.

Male flower
X 0.5

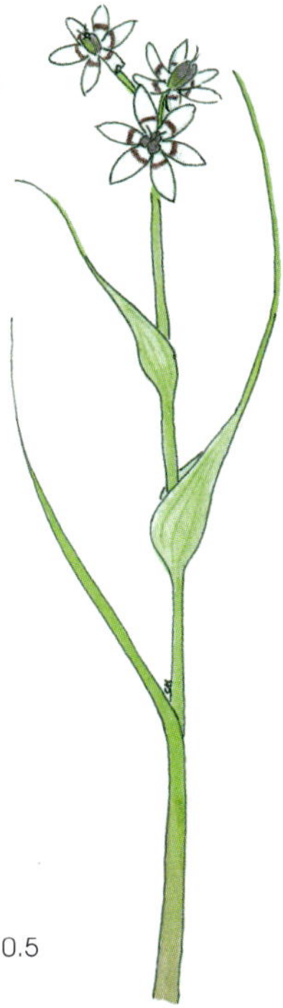

X 0.5

Flowering Period

Late winter and early spring.

Distribution

Forms extensive colonies in dry woodlands in the east of the State.

Locality

Hillside above Howrah on Hobart's eastern shore. Tunbridge tip.

Cultivation Notes

Probably can be grown from seed.

Related Species

W. unifolia is the other reasonably common species in this genus in Tasmania. It has only one or two flowers, which are bisexual and have yellow anthers. It is usually found on peaty wet ground including button grass plains, mostly in the north of the State. *W. latifolia* is localised in the north-west of the State. Closely related and common is *Burchardia umbellata* which is a much larger lily with a cluster of white flowers at the end of its main stem.

Diplarrena moraea

Iridaceae

White Iris

X 0.2

Distinguishing Features

A tufted plant with leathery flat leaves. White iris-like flowers emerge from the top of long flat stems similar to the leaves.

Description

When not in flower this plant looks similar to many of the sedges and saggs which grow in similar habitats. The finely striped leaves which end in a single point distinguish this species from the others. After the flowers have finished, the three-lobed cylindrical capsules may be found at the ends of the old flower stems.

Flower
X 0.5

Flowering Period

Late spring and early summer.

Distribution

Common in dry woodlands.

Locality

Mt Nelson, south of Hobart.

Cultivation Notes

Propagation is by division or from seed. Flowers do not last long in the garden.

Related Species

D. latifolia is a very similar species from wet moorland in the west and alpine habitats. The leaves of this species tend to be wider than in *D. moraea*. There is an outstanding display of *D. latifolia* each summer near the King William Saddle beside the Lyell Highway. Other closely related species include the blue irises *Patersonia fragilis* and *P. occidentalis*. Both of these species grow in wet heaths in the lowlands.

Xanthorrhoea australis

Xanthorrhoeaceae

Grass Tree

Distinguishing Features

A plant with a large rounded crown of long linear leaves, on a trunk which can be up to 2m tall.

Description

When present the trunk is usually blackened by fires. It is these fires which promote the flowering which may be in the form of a spectacular display with hundreds of the tall flower spikes present together. These spikes are packed with flowers which attract a large variety of nectar-loving insects. White stamens are the prominent feature of the flowers. The dead flower spikes may persist for several years.

Flowering Period

Mainly in spring after a recent fire.

Distribution

Local near the coast in northern and eastern parts of the State.

Locality

Rocky Cape National Park. Roadside at Killora, Bruny Island. Freycinet National Park.

Cultivation Notes

Can be grown from seed, but it grows very slowly.

X 0.1

Related Species

Tasmania has two main types of grass tree. The larger one is described here, while the smaller one never has a trunk, has multiple tufts of leaves at ground level and has a slender flowering stem. The smaller form has recently been divided into two species, *X. bracteata* and *X. arenaria*. Distinguishing between these species is difficult. All of the 28 species of *Xanthorrhoea* are restricted to Australia. They form a most distinctive part of our flora.

Kunzea ambigua
Myrtaceae

Distinguishing Features
A dense spreading shrub with a profusion of white flowers with small hidden petals and numerous prominent stamens.

Description
This is one of the many heathland plants with narrow linear leaves about 1cm long. In this case the crushed leaves smell of eucalyptus oils and the shrub can be up to 3m tall. The copious flowers with their sweet scent make a fine display. Later the clustered capsules can be seen but these shed their seed and drop off before winter.

Flowering Period
Late spring.

Distribution
Sandy heathland on the east coast, north from Freycinet Peninsula.

Locality
Car park at the start of the Wineglass Bay Track, Freycinet National Park. Mt William National Park.

X 1.0

Cultivation Notes
Very easy to propagate from cuttings. Grows fairly quickly into a large shrub. Responds to pruning.

Related Species
There are several species of *Kunzea* in southern Australia. Many of the mainland species have attractive yellow or pink-purple flowers. At Kosciusko the yellow kunzea, *K. muelleri,* puts on a magnificent display of yellow in midsummer. In Tasmania *K. ambigua* most closely resembles *Calytrix tetragona* (illustrated on p.15).

Leptospermum scoparium

Tea-tree

Myrtaceae

Distinguishing Features

A shrub which is usually slender and upright. The leaves are hard, shiny and sharply pointed.

Description

A member of the tea-tree genus which has white flowers with five round petals. At the base of the petals is a ring of short stamens. The rounded capsules have five prominent lobes on top. Sometimes plants are affected by red galls caused by insect attack.

Flowering Period

Summer.

Distribution

Very common, particularly near the coasts.

Locality

Start of the Wineglass Bay Track, Freycinet National Park. Peter Murrell Reserve near Blackmans Bay.

Cultivation Notes

Easily propagated from seed.

Related Species

The eight species of *Leptospermum* in Tasmania are common and locally abundant. Several species favour dry sites in the east of the State. *L. grandiflorum* has large flowers and fruits with grey-green leaves. It is restricted to places near the east coast. Another species with grey-green leaves is *L. glaucescens*. This species has purple fruits which are less woody than the fruits of other species. In damper sites the woolly tea-tree, *L. lanigerum*, is common. The woolly parts are the hairs on the bases of the leaves and on the fruits. All of the species mentioned may be found in Freycinet National Park.

X 1.0

Eucalyptus pulchella
Myrtaceae

White Peppermint

Distinguishing Features
A graceful tree with white or pale-grey bark. The leaves are long and narrow.

Description
A small to medium tree with a short trunk. Sometimes there is a little rough bark around the base but otherwise it is smooth throughout. The flowers and fruit are clustered into groups of up to 12. The typical gum flowers have massed stamens and no obvious petals.

Flowering Period
Mainly in late spring, but also at other times of the year.

Distribution
Mostly on fertile well-drained hillsides in the south-east.

Locality
North-facing slopes in Ridgeway Park, near Hobart. Beside Hobart's Southern Outlet.

Cultivation Notes
Easily propagated from seed. Grows reasonably fast.

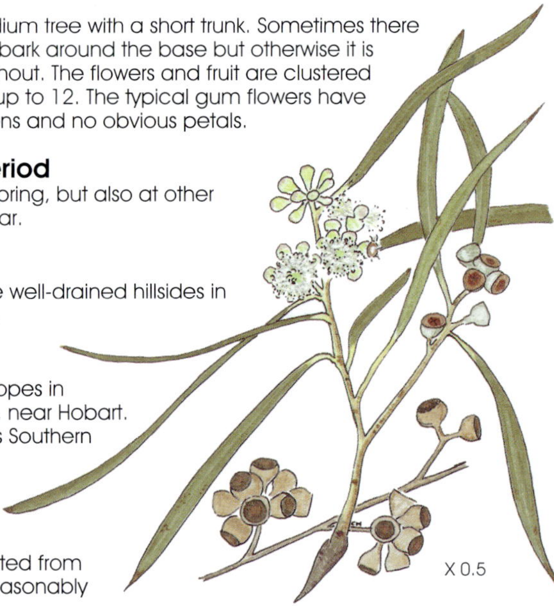

Related Species
About twenty different species of *Eucalyptus* trees are found in lowland woodlands, many being more or less specialised to certain habitats. *E. amygdalina,* black peppermint, is closely related to *E. pulchella.* Its main difference is in the bark which is rough and fibrous. In some localities there are trees with rough bark halfway up the trunk which are intermediate between these species. *E. amygdalina* is very common in Tasmania, especially on poor soils.

Calytrix tetragona
Myrtaceae

Fringe Myrtle

Distinguishing Features

The white flowers tend to be densely clustered. Each flower has five long filaments growing out from below the petals. There are many prominent stamens.

Young fruit X 0.5

Description

One of a large number of heathland shrubs which have narrow linear leaves about 1cm long. This shrub is in the same family as eucalyptus trees, and the leaves contain various oils which give off a scent when crushed. It is rarely more than 1m tall, and frequently has branches which spread out widely from the base.

X 0.5

Flowering Period

Mainly spring and summer.

Distribution

Mostly restricted to heathlands near the coast.

Locality

At the start of the track to Wineglass Bay, Freycinet National Park. First Basin at the Cataract Gorge, Launceston.

Cultivation Notes

May be propagated from seed or cuttings.

Related Species

The stronghold of *Calytrix* is in Western Australia where there are many species with pink to mauve petals and contrasting yellow stamens. *C. tetragona* is currently listed for all Australian states, but is very variable particularly in the hairiness of the leaves and in the oils found in the leaves.

Ricinocarpos pinifolius
Euphorbiaceae

Wedding Bush

Distinguishing Features

A shrub with narrow linear leaves and attractive white flowers on long stems.

Description

A densely branched shrub about 1m tall. The leaves are usually more than 1cm long. They are dark green above and have edges which are curved back and partly obscure the pale lower surface. Each flower is either male or female. Female flowers are fatter than the male flowers.

X 1.0

Flowering Period

Mostly in spring.

Distribution

Locally common near the coast in the north-east.

Locality

Bridport. Eddystone Point, Mt William National Park. Approaches to Coles Bay.

Cultivation Notes

Highly desirable, but not easy to propagate and grow.

Related Species

The family Euphorbiaceae is very large and diverse. There are about 7500 different species with plants varying from the garden weed, sun spurge, through the wedding bush with its pretty flowers, to fleshy cactus-like plants in Africa. Many species have a milky latex which may be seen when a stem is broken. Commercial use is made of the latex found in a Brazilian species which is the source of rubber. Tasmania shares a diverse flora in this family, including the almost leafless *Amperea xiphoclada* which is common on poor soils. This has tiny flowers, as do all of the other Tasmanian species in the family except the wedding bush.

Stackhousia monogyna

Stackhousiaceae

Candles, Native Mignonette

Distinguishing Features

A multi-stemmed herb with thick linear leaves and a dense spike of white flowers. Each flower has a slender tube and spreading petals.

Description

A variable herb which can be very fleshy with glossy green leaves, or much more slender and elegant. The white flower spike which resembles a candle is usually sweetly scented. Three little round fruits cling to the stem after each flower falls.

Flowering Period

Mainly in spring in the lowlands through to early autumn in the highlands.

Distribution

Widespread and extending well into the mountains. Often prominent after fires.

Locality

Cape Hauy, especially on shady cliff edges. State forest south of Mt Hobbs, which is north-west of Buckland.

Cultivation Notes

Easily grown from seed. Lasts for several years, but best treated as short-lived.

X 0.5

Related Species

There are five species of *Stackhousia* in Tasmania. All of the other Tasmanian species are uncommon and restricted to specific habitats. A single-stemmed variant of *S. monogyna* which was previously called *S. gunnii* can be seen in the Midlands at the Tunbridge tip and in the Campbell Town/ Conara area. *S. spathulata* has larger leaves than *S. monogyna* and grows mainly on sand-dunes in the west and south. The two other species have creamy-yellow flowers.

Pimelea linifolia
Thymelaeaceae

Rice Flower

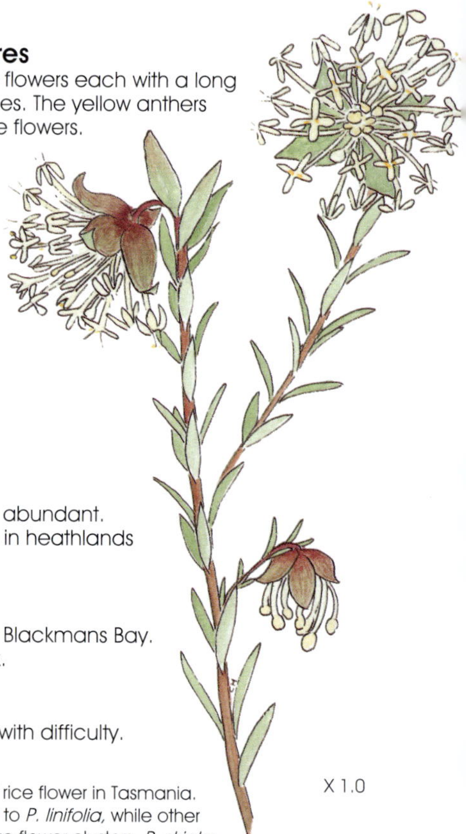
X 1.0

Distinguishing Features
A drooping head of white flowers each with a long narrow tube and short lobes. The yellow anthers are prominent beyond the flowers.

Description
An erect shrub up to about 1m tall, with few branches. The grey-green leaves are opposite each other along the stem. Flower heads are surrounded by four modified leaves.

Flowering Period
Spring.

Distribution
Widespread but not often abundant. Mostly found on poor soils in heathlands or woodlands.

Locality
Peter Murrell Reserve near Blackmans Bay.
Rocky Cape National Park.

Cultivation Notes
Propagates from cuttings with difficulty.

Related Species
There are eighteen species of rice flower in Tasmania. Some of these are very similar to *P. linifolia,* while other species do not have such large flower clusters. *P. stricta* from the east coast is the most similar, differing only in the detail of the hairs on the flowers and usually having the flower clusters upright and not drooping. *P. humilis* is a very low-growing species common in dry places. This has green leaves and hairy stems. *P. glauca* which is common on sandy dunes in the east has grey leaves which are densely packed along the stem.

Philotheca verrucosa

Wax Flower

(formerly Eriostemon verrucosus)
Rutaceae

Distinguishing Features

Grey-green shrub with prominent warts on the leaves and stems.

Description

Often an untidy shrub with weak trailing branches up to 50cm long. The leaves are thick and narrowed at the base. The star-shaped flowers are typically pink in bud and white when fully open. Occasional pink-flowered plants may be found.

Flowering Period

Late winter and spring.

Distribution

Usually on dry rocky sites, sometimes on river banks.

Locality

East Risdon Nature Reserve on Hobart's eastern shore.
Freycinet National Park.
Roadsides from Cranbrook to Avoca and between Launceston and Underwood.

Cultivation Notes

Difficult to propagate from cuttings. There are several named cultivars most of which originate from Victoria.

X 1.0

Related Species

An easy way to recognise a boronia is to notice that it has only four petals. *Philotheca* is similar but the flowers normally have five petals. Two other species of *Philotheca* are found in Tasmania. *P. virgata* is a larger shrub with no warts. It is locally common in Freycinet National Park, Rocky Cape National Park and near the west coast. *P. freyciana* is a newly described species growing on granite in the Freycinet area.

Ozothamnus scutellifolius

(formerly Helichrysum scutellifolium)
Asteraceae

Distinguishing Features

Multi-branched shrub with white woolly stems and tiny dark-green leaves.

Description

The small flower heads are closely packed and each is surrounded by small papery bracts. Plants are usually untidy and less than 1m tall.

Flowering Period

Late spring and early summer.

Distribution

Dry rocky places in the south-east of the State.

Locality

Lake Leake Road about 20km west of Swansea.

Cultivation Notes

Easy to strike from cuttings. Grows well in many soils.

Related Species

The genus *Ozothamnus* represents the shrubby daisy bushes formerly included in *Helichrysum*. Many of them are found in the woodlands in the east of the State. *O. ferrugineus* (formerly *H. dendroideum*) is a large shrub with linear leaves and dense clusters of small flowers at the ends of many branches. It is similar to several other related species. *O. obcordatus* is different from most other shrubby species as it has dense clusters of yellow flowers and small rounded leaves.

Bursaria spinosa ssp. **spinosa** Prickly Box
Pittosporaceae

X 1.0

Old capsules X 1.0

Distinguishing Features
A bushy upright shrub or small tree with spines amongst the thin dark-green leaves. The prominent white flower clusters are at the ends of the branches.

Description
This shrub tends to have a gnarled appearance with its scaly bark. The dry capsules tend to persist on the plant long after the seeds have fallen.

Flowering Period
Mostly in mid-summer.

Distribution
Very common in dry stony places in the east and north. Often a remnant in paddocks which are otherwise cleared.

Locality
Bakers Beach, Narawntapu National Park. Start of the track to Cape Hauy.

Cultivation Notes
Easy to grow from cuttings and may even appear voluntarily in gardens from wild seed. It is adaptable, but prefers dry sites.

Related Species
There is no other species of *Bursaria* in Tasmania but there is considerable variation in leaf size and spininess of the plants. The genus *Bursaria* has eight species and is confined to Australia.

Clematis gentianoides

Ranunculaceae

Distinguishing Features

Small plant with many leafy stems arising in a low tuft. Relatively large white flowers at the ends of the stems.

Description

This is the only non-climbing species of *Clematis* in Tasmania. The leaves are dark green with veins clearly visible. Usually they are undivided, but are occasionally serrated. Male and female flowers occur on separate plants. Flowers have a variable number of petal-like sepals. Female flowers are followed by feathery seeds.

Flowering Period

Late spring and early summer.

Distribution

Restricted to dry rocky places in the east between Hobart and Bicheno and near Launceston.

Locality

Waverley Park Nature Reserve at Bellerive. Kate Reed Reserve, Launceston.

Cultivation Notes

Easily grown from seed.

Related Species

The two other native species of *Clematis* are climbing plants like the popular cultivated varieties. They both have white flowers and divided leaves. The common *C. aristata* has leaves divided into three stalked leaflets. *C. microphylla* is restricted to sand-dunes near the north coast, and this has its leaves divided into three parts which are themselves divided into three leaflets.

Lomatia tinctoria
Proteaceae

Guitar Plant

Distinguishing Features
The leaves appear fernlike, but are leathery to the touch.

Description
A shrub which is usually seen in extensive colonies, as it spreads by underground rhizomes. Above ground it appears as a plant with few stems and is rarely more than 1 m tall. The white or cream flowers are in loose spikes which are followed by guitar-shaped woody fruits.

Old fruit
X 0.5

Flowering Period
Summer.

Distribution
Widespread and often abundant except near the west coast and in the Midlands.

Locality
Wineglass Bay Track, Freycinet National Park. On the track to Cape Hauy where there is extreme variation in leaf shape.

Cultivation Notes
May be propagated from seed or cuttings. May benefit from pruning.

Related Species
The other two members of the genus *Lomatia* in Tasmania are not found in woodlands. *L. polymorpha* is a plant of the mountains. It has leaves which are less divided than those of the guitar plant and the flowers are densely clustered. Tasmania's other species, *L. tasmanica*, is known only from a small area of rainforest near Port Davey. It has red flowers. Apparently it has lost the ability to regenerate from seed as new plants form only as suckers from existing plants.

X 0.5

Leucopogon collinus

Epacridaceae

Bearded Heath

Distinguishing Features

Clusters of tiny white hairy flowers at the ends of branches. The flowers are closely surrounded by brown sepals.

Description

A small shrub which is usually under 50cm tall. The leaves have their edges curved back and end in a point. They are usually grey-green and much shorter than 10mm long.

X 1.0

Flowering Period

All year.

Distribution

Abundant in many places especially in sandy soils. A prostrate form is common on infertile mountains in the west.

Locality

Hazards Beach Track west of Mt Mayson, Freycinet National Park. Rocky Cape National Park. Peter Murrell Reserve near Blackmans Bay.

Cultivation Notes

Can be propagated from cuttings. Probably best used as a rockery plant.

Related Species

All 13 species of *Leucopogon* in the State have white flowers which are more or less hairy. *L. ericoides* is very similar to *L. collinus*, but it differs by being a taller shrub with flower clusters at the bases of the leaves. Each flower is surrounded by white sepals. Another species encountered frequently in dry places is *L. virgatus*. This has leaves with edges curving upwards and flower clusters towards the ends of the branches.

Epacris tasmanica
Epacridaceae

Distinguishing Features
Long leafy spikes of white flowers on a shrub with hard, sharply pointed leaves which have a short stalk.

Description
A shrub with few branches, usually under 1.5m tall. The floral tube is hidden in the hard bracts surrounding the flower, but the showy petals surround the stamens and style which are a prominent feature of the flower.

Flowering Period
Late spring.

Distribution
Abundant in several localities in the east and south-east.

Locality
Cherry Tree Hill, Tasman Highway 20 km south-west of Bicheno.

Cultivation Notes
May be propagated from cuttings or seed. Pruning will encourage a more bushy shrub.

X 1.0

Related Species
(See also *E. impressa* p. 30 in this booklet.) There is a large group of epacris which looks similar to *E. tasmanica*. They are separated mainly by features of the carpel and stamens. *E. tasmanica* has been divided, with the smaller leafed form growing south of Hobart now being included in *E. virgata*. Some other species are very local in distribution, although they may be abundant in their habitat. Of these *E. barbata* may be encountered in Freycinet National Park. This has hairs on the bracts surrounding the flowers.

Euryomyrtus ramosissima

(formerly Baeckea ramosissima)
Myrtaceae

Distinguishing Features

A wiry and prostrate shrub with small opposite leaves. The flowers are on long stems and have round petals that are usually pale pink.

Description

This variable shrub has leaves which are between 4mm and 10mm long. The flowers have only ten stamens in contrast to the similar-looking tea-tree flowers which have many stamens. Often a few petals are blotched with darker pink.

Flowering Period

Mainly late spring, but a few flowers may be found in winter.

Distribution

Sandy heathlands and woodlands throughout the State. Also locally abundant on rocky river banks.

Locality

Archers Knob, Narawntapu National Park. Peter Murrell Reserve near Blackmans Bay.
W. M. Curtis Reserve, Scamander.

Cultivation Notes

Fairly easy to propagate from cuttings. Responds well to pruning, but liable to die suddenly.

X 1.0

Related Species

There are no other species of *Euryomyrtus* in Tasmania. However *E. ramosissima* has two forms ssp. *ramosissima*, which is more shrubby and occurs in the east and south-east, and ssp. *prostrata* which occurs in the north of the State and has smaller flowers than the illustration. Similar small flowers occur on *Baeckea leptocaulis* from peaty heathlands in the west of the State and *B. gunniana* which is common in boggy alpine situations. *Thryptomene micrantha* which is abundant in Freycinet National Park is a spreading shrub with arching branches, crowded rounded leaves and small white flowers which turn pink as they age.

Drosera peltata ssp. **auriculata** Sundew

Droseraceae

Distinguishing Features

The small shield-shaped leaves are covered with stout hairs called tentacles, which bear a blob of sticky liquid at the tip. No tentacles are attached to the base of the flowers.

Description

A perennial herb which dies back to ground level in summer. The sticky fluid on the tentacles may capture small insects. Enzymes in the fluid then partially dissolve the body which provides nutrients for the plant. The pink flowers appear in a small group at the end of the stem. In large plants the stem can be branched.

Flowering Period

Spring.

Distribution

Common in the lowlands. Often found in thin soil on rocks, in dry forest, and sometimes in peaty ground.

Locality

Heathlands at Bridport and at Cloudy Bay on Bruny Island.

Cultivation Notes

Seed may be sown in spring. Requires lime-free infertile soil. Tubers transplanted from the bush usually fade away.

Related Species

Tasmania has seven species of *Drosera*, all of which are armed with sticky droplets on their tentacles. Two other species may be confused with *D. peltata* ssp. *auriculata*. *D. peltata* ssp. *peltata* has each flower surrounded by a ring of hairy sepals. *D. macrantha* is a climbing sundew with small round leaves and relatively larger flowers. This is found mostly in sandy woodlands and heaths in the north of the State.

Boronia pilosa
Rutaceae

Distinguishing Features
The leaves of this shrub are opposite each other, are divided into linear segments and have a mildly unpleasant smell when crushed.

Description
This small variable shrub is usually about 50-75cm tall with erect stems. The leaves may be hairy, as the scientific name suggests, but more often they are not hairy. The pink or white flowers are crowded at the ends of the branches. Each flower has only four petals.

X 1.0

Flowering Period
Spring.

Distribution
Locally abundant. Widespread throughout the lowlands.

Locality
Beside the track southwest of Banksia Grove, Rocky Cape National Park.
Freycinet National Park.
W. M. Curtis Reserve, Scamander.

Cultivation Notes
Propagation from cuttings is not easy but is worth the effort. May be short lived in cultivation, but pruning and partial shade may extend its life.

Related Species
B. pilosa is very similar to *B. citriodora* which favours alpine parts of the State. The easy way to tell these two species apart in the wild is to crush a leaf. *B. citriodora* has a pleasant lemon scent, unlike the less pleasant smell from *B. pilosa*. The most unpleasant smelling boronia in Tasmania is *B. anemonifolia*. A crushed leaf of this species has a turpentine-like smell. *B. anemonifolia* often favours dry infertile hillsides where it grows up to 2m in height. There are also three species of *Boronia* in Tasmania which are dwarf shrubs with undivided leaves.

Lissanthe strigosa ssp. **subulata** Peach Berry

(formerly Lissanthe strigosa)
Epacridaceae

Distinguishing Features

A shrub which often forms a low-growing thicket. The 10mm-long leaves are narrow-linear and sharply pointed with white stripes underneath.

Description

A densely branched shrub which is usually under 20cm tall. The flowers are typically white or pale pink with the buds being a deeper pink. The fleshy fruits are white or pale pink when ripe, like miniature peaches, but do not seem to remain on the plant for very long.

Flowering Period

Late spring.

Distribution

Common. Usually in the understorey of woodlands in dry places.

Locality

Queens Domain and Ridgeway Park near Hobart. Road verges south of Cressy in the northern Midlands.

Cultivation Notes

Can be propagated from cuttings or seeds with difficulty.

Related Species

There is no other species of *Lissanthe* in Tasmania. *Leucopogon montanus* was formerly included in *Lissanthe*. It is a small alpine shrub with blunt leaf tips.

Epacris impressa

Epacridaceae

Common Heath

Distinguishing Features

Showy red, pink or white tubular flowers on a shrub with hard sharp leaves.

Description

A very variable shrub with few branches, usually under 1m tall. The flower tube has a number of impressions around the base. Flowers are one of the variable parts of the plant. In some localities they are about the same length as the leaves.

Flowering Period

All year.

Distribution

Common in many situations from sandy heathlands to well-drained sites on quartzite mountains in the west.

Locality

Some of the best specimens may be found at Murdunna on the Tasman Peninsula, and at the Dial Range in the north-west.

Pink and red flowered forms
X 1.0

Cultivation Notes

Difficult to propagate from cuttings and from seed. Seed is said to germinate better while fresh.

Related Species

(See also *E. tasmanica* p. 25 in this booklet.) There are about 24 species of *Epacris* in Tasmania, and most are restricted to the lowlands. There has been active research into this genus for many years. Consequently it is difficult for an amateur to put a name to many plants that are found. *E. obtusifolia* and *E. lanuginosa* are the other species with a floral tube which is longer than the sepals. Both favour damp peaty heaths and have white flowers. The leaves of *E. obtusifolia* are blunt while those of *E. lanuginosa* are narrow and pointed.

Caladenia clavigera Clubbed Spider Orchid

Orchidaceae

X 0.5

Distinguishing Features

A ground orchid which has one or two pale-red spider-like flowers at the end of an erect stem. The labellum, in the centre of the flower, has plain margins (not fringed with stout hairs).

Description

A perennial herb which survives the summer as an underground tuber. The hairy stem has one strap-like hairy leaf growing adjacent to its base. The ends of the elongated petals and sepals sometimes have tiny clubs, which is the origin of the orchid's name.

Flowering Period

Spring.

Distribution

Local in woodlands especially in dry places in scattered eastern localities.

Locality

Hillsides north of Howrah on Hobart's eastern shore. Bridport.

Cultivation Notes

Orchid propagation is a specialised field. It may be possible to buy dormant tubers from specialised nurseries.

Related Species

In Tasmania there are more than 20 species of *Caladenia* which form two distinct groups. *C. clavigera* is a member of the group called spider orchids. The most frequently seen spider orchids are *C. caudata* (north, east and south), *C. dilatata* (north) and *C. echidnachila* (south).

Pink fingers, *C. carnea,* is a common member of the group of smaller caladenias. This forms extensive patches in some localities. Also common on dry inland sites is the white-flowered *C. angustata* which has a hood over the labellum and often smells of musk.

Astroloma humifusum

Epacridaceae

Native Cranberry

Distinguishing Features

A prostrate shrub which forms a small mat. The red tubular flowers are usually inconspicuous, being partly hidden by the leaves.

Description

This very prickly shrub is easily overlooked. The leaves may be as long as 1cm with a long point at the end. The edible fleshy fruit is up to 1cm in diameter with a relatively large stone inside.

Flowering Period

Mainly in the autumn.

Distribution

Common on dry rocky sites in woodlands and in sandy heathlands.

Locality

Badger Head, Narawntapu National Park. North-facing slopes in Ridgeway Park, near Hobart.

Cultivation Notes

Rather difficult to strike from cuttings, and probably also difficult to germinate from the woody fruit.

X 1.0

Related Species

The other Tasmanian member of this genus is the uncommon *A. pinifolium* (see Coastal Plants Identikit) which is a very attractive pine-like small shrub. The yellow flower tubes are about 15mm long and easily seen. It often favours sandstone ledges as at Buckland Gorge near Orford. Two other members of the heath family are very similar to the species of *Astroloma*. *Styphelia adscendens* is described in this booklet. Ant's delight, *Acrotriche serrulata*, has smaller leaves than this species and green flowers in clusters hidden by the prickly foliage.

Kennedia prostrata
Fabaceae

Running Postman

Distinguishing Features
A prostrate herb with large clover-like leaves and bright red flowers.

Description
A member of the pea family which is very distinctive when in flower. The flowers are mainly red with two small yellow-green eyes near the centre. The prostrate branches can be nearly 1m in length. The small black seeds, which are contained in a long brown pod, have a white attachment.

Flowering Period
Spring.

Distribution
Widespread in dry or sandy soils especially near the coast, but rarely common.

Locality
Eaglehawk Neck, Tasman Peninsula. Stieglitz near St Helens.

Cultivation Notes
Easily grown from seed after soaking in boiling water. Survives for several years before needing to be resown.

Related Species
This distinctive plant has no close relations in Tasmania. *K. prostrata* is found throughout Australia, and is especially prolific in Western Australia, with leaves and flowers much bigger than can be found in Tasmania. Several species are often cultivated in Tasmania, including *K. nigricans* which is a vigorous climber with black flowers. *K. coccinea* is also a climber with spectacular clusters of red flowers. Both of these cultivated species occur naturally in Western Australia.

Cyphanthera tasmanica
Solanaceae

Distinguishing Features
A grey-green shrub which is more or less densely covered with tiny hairs. The showy flowers are creamy-white with purple stripes in the centre.

X 1.0

Description
This is an unusual member of the potato family. It varies from a lush fleshy shrub in protected sites to a harsh scruffy shrub in exposed coastal sites. The leaves have a rough sandpaper-like feel because of microscopic hairs.

Flowering Period
Spring.

Distribution
Rare, but occasionally abundant as at Mt Puzzler near the Meadstone Falls. Mostly within 30km of the east coast. Usually on rocky sites.

Locality
West of Mt Mayson on the Hazards Beach Track, Freycinet National Park.

Cultivation Notes
Easy to grow from cuttings but forms into a straggly shrub unless it is carefully pruned.

Related Species
This species is restricted to Tasmania. Until recently it was known as *Anthocercis tasmanica*. All nine species of *Cyphanthera* are found in southern Australia, but no others extend to Tasmania. They all have similar flowers with prominent purple stripes. Several species appear to be much more common after fires, sometimes growing as dense thickets. Very few members of the potato family are native to Tasmania. Probably the most common is the kangaroo apple, *Solanum laciniatum*.

Leptecophylla divaricata

(formerly Cyathodes divaricata)
Epacridaceae

Distinguishing Features

Very prickly narrow leaves which are striped white underneath. The small cream flowers have a short stalk.

Flower X 1.0

Description

This shrub is usually about 1m tall. The narrow linear leaves are often crowded, particularly at the ends of the branches. The fruits vary in colour from pale-pink to red and are about 8mm in diameter.

Flowering Period

Late autumn to early spring.

Distribution

Dry rocky sites in woodlands usually within 40km of the south-eastern coast.

Locality

Chimney Pot Hill, near Hobart.
Tasman Highway west of Orford.

Cultivation Notes

Difficult to propagate from cuttings and from seed, which are said to germinate better after passing through a bird. Not commonly cultivated.

Fruit X 1.0

Related Species

A number of species formerly included in *Cyathodes* have been separated into the genera *Planocarpa* and *Leptecophylla*. *L. pendulosa* is found on rocky upland sites between Launceston and Swansea. It differs from *L. divaricata* by having white flowers usually alone at the end of a drooping branch. *L. juniperina* is widespread from sea level to mountain summits with three subspecies described. There are four species of *Cyathodes* and three of *Planocarpa*. Most species in this group have attractive pink or red fruits.

Styphelia adscendens
Epacridaceae

Distinguishing Features

The greenish-yellow tubular flowers are fairly conspicuous with at least the hairy lobes of the petals showing through the mat of hard pointed leaves.

Description

A prostrate shrub which forms a mat of flat leaves. The leaves are relatively large for a member of the heath family. They are dark green above and paler below with fine veins clearly visible. The fleshy fruit is also relatively large with a large stone inside.

Flowering Period

Mainly in winter, less commonly in spring.

Distribution

Scattered in districts mainly near the coast. Sometimes on dry sites, but also on damp peaty sites.

Locality

Damp flats on the track to Point Vision, Narawntapu National Park. Coningham State Recreation Area south of Hobart.

X 1.0

Cultivation Notes

May be grown from cuttings, but probably difficult to germinate from the woody fruit.

Related Species

No other member of this genus occurs in Tasmania. It is most likely to be confused with *Astroloma humifusum* which is described in this booklet (p. 32). A. *humifusum* has leaves about half the size of this species and has red flowers. Other members of the heath family which have fleshy fruit are *Leptecophylla divaricata* (p. 35), *Lissanthe strigosa* ssp. *subulata* (p. 29) and *Leucopogon collinus* (p. 24).

Hakea epiglottis
Proteaceae

Distinguishing Features
The leaves are sharp, rigid, well-spaced needles which curve upwards near the base.

Description
A large shrub up to 5m tall, but typically less than this. Flowers are in clusters at the bases of the needles. They vary from white through creamy yellow to bright yellow. The woody fruit has a beak which is curved upwards.

Flowering Period
Some plants near the east coast flower in autumn. Elsewhere flowers occur in spring.

Distribution
Dry rocky sites near the east coast. Damp heaths up to 1000m elsewhere.

Locality
North-west of Mt Mayson on the Hazards Beach Track, Freycinet National Park. Track to Cape Hauy.

Cultivation Notes
When collected, the woody fruits soon open to release the seeds, which are easy to germinate.

X 1.0

Related Species
There are several species of *Hakea* in Tasmania. *H. lissosperma* is usually found in mountainous areas, but is also found on the Wineglass Bay Track, Freycinet National Park. It has long straight needle-like leaves and showy white flowers. *H. teretifolia* is common in coastal heathlands, especially in damp sites. It is a very tangled bush with straight spines and a pointed woody fruit. *H. microcarpa* is often found on rocky river banks in the east. As the name implies this has small woody fruits and these open on the bush when mature.

Banksia serrata

Proteaceae

Saw Banksia

Distinguishing Features

Usually a small tree with thick wrinkled bark on the trunk. The persistent old cones contain several large rounded woody fruits which are clearly visible.

Description

The hard shiny leaves of this gnarled tree have a serrated edge. They are shiny on top but less shiny below because of numerous tiny pits in the surface. Being at the ends of branches, the flower spikes are very prominent.

Flowering Period

Summer and early autumn.

Distribution

Mostly in Rocky Cape National Park.

Locality

Banksia Grove, at the western end of Sisters Beach, Rocky Cape National Park. Also in the township of Sisters Beach.

Cultivation Notes

Easy to grow from seed. It can be a slow-growing plant which will take a number of years to flower.

X 0.5

Related Species

The only other species of banksia in Tasmania is *B. marginata*. Flower spikes are smaller than in *B. serrata* and the leaves are much less serrated. It is common from seashores to high in the mountains.

Gompholobium huegelii

Bladder Pea

Fabaceae

Distinguishing Features

Pea flowers with petals which are creamy-yellow inside and dark olive green on the outside.

Description

A multi-stemmed low-growing shrub with narrow grey-green leaves which appear to be in clusters along the stem. The oval buds are like small clubs on a long stem. The pods are a similar shape but slightly larger.

Flowering Period

Late spring and summer.

Distribution

Common, but rarely abundant, in sandy heathlands all around the State. Also found in dry woodlands.

Locality

Rocky Cape National Park. Near the Ida Bay Railway at Southport. Behind Hazards Beach, Freycinet National Park. Peter Murrell Reserve near Blackmans Bay.

Cultivation Notes

Easily germinated from seed after soaking in boiling water. May be difficult to grow successfully.

Related Species

The other species of *Gompholobium* recorded from Tasmania is only found on Flinders Island. This is *G. ecostatum*, which has red flowers. Like many of Australia's pretty wildflowers the stronghold of this genus is in Western Australia with about 16 species.

Pomaderris elliptica

Yellow Dogwood

Rhamnaceae

Distinguishing Features

A large shrub with large soft leaves which are shiny above but white below because of a dense covering of microscopic hairs.

Description

The yellow flower heads of this species are very attractive. The flower buds are also bright yellow during the winter months. Each flower is small with prominent stamens. The soft leaves resemble those of a deciduous tree rather than the hard leaves typically found on plants in the drier parts of the State.

Flowering Period

Late spring.

Distribution

Mostly near the east and north coasts, but occasional near the west coast and inland.

Locality

Along the track to Cape Hauy. Narawntapu National Park.

Cultivation Notes

May be propagated from cuttings with difficulty, or seed which responds to pre-sowing treatment. Highly desirable.

X 0.5

Related Species

Plants in the genus *Pomaderris* are often difficult to identify without a microscope. There are about four species which look identical to *P. elliptica* without this careful study. *P. pilifera* has small brown hairs along the mid-rib of the underside of the leaf. This species seems to favour inland localities such as the Cataract Gorge in Launceston. Flinders Island is the main home for *P. intermedia* with its cocktail of different types of hairs on the lower leaf surface.

Melaleuca squarrosa

Scented Paper-bark

Myrtaceae

Distinguishing Features

Small bottle brush-like flowers which are usually pale yellow. The crowded leaves are about 10mm long and are opposite each other along the stem.

Description

A shrub which may be densely branched or slender and growing in a dense thicket. The leaves are rounded at the base, and end in a soft point. When crushed the leaves are scented. The fruits are small woody capsules which are closely packed in spikes.

Flowering Period

Late spring and early summer.

Distribution

Common in shallow gullies and damp areas in heathlands and forests.

Locality

Damp places along the track to Cape Hauy and in Rocky Cape National Park. Peter Murrell Reserve, near Blackmans Bay.

Cultivation Notes

May be propagated from seed or cuttings. Suited to damp sites in the garden.

Related Species

There are five species of *Melaleuca* in Tasmania. The only one to favour dry sites is *M. pustulata* which is restricted to a small part of Tasmania. It may be seen at Cherry Tree Hill on the Tasman Highway between Swansea and Bicheno. *M. ericifolia* has similar leaves and flowers, but forms large copses in wet areas in the north and east of the State. The white paper-bark on the trunk of *M. ericifolia* is very noticeable.

X 1.0

Callistemon pallidus
Myrtaceae

Yellow Bottlebrush

X 0.5

Distinguishing Features

The large bottlebrush flowers which are creamy-yellow to lemon-green in colour.

Description

A shrub which is usually upright with pale bark. The leaves are about 5cm long. They are hard with a clear midrib and with obvious dots in the leaf. These are glands which give off a typical eucalyptus-like smell when a leaf is crushed. The fruit is a woody capsule which persists in spikes along the smaller branches.

Flowering Period

Late spring and early summer.

Distribution

Common on rocky river banks and dry stony hillsides in the east.

Locality

Prosser River, Orford. Western side of the First Basin at the Cataract Gorge, Launceston.

Cultivation Notes

Easily grown from seed. May be pruned after flowering. Adaptable in the garden.

Related Species

There is one other common species of *Callistemon* in Tasmania. This is *C. viridiflorus* which grows in marshes and damp habitats mostly above 500m in altitude. It has smaller bright-green leaves and smaller flower spikes than *C. pallidus*. Species of *Melaleuca* also possess bottlebrush flowers which are usually smaller than those of Callistemon (see *M. squarrosa* p. 41).

Acacia suaveolens
Mimosaceae

Sweet Wattle

Distinguishing Features
The flower buds are enclosed in brown overlapping scales. Round clusters of flowers later emerge on short branches.

Description
A straggly shrub with few branches, usually less than 1m tall. The phyllodes (leaf stems which resemble leaves) are grey-green, leathery and up to 15cm long. The thickened stems at the leaf bases are also distinctive.

Flowering Period
Late autumn and winter.

Distribution
Common on sandy heathlands especially near the coast.

Locality
Heathlands in Rocky Cape National Park and in Freycinet National Park.

Cultivation Notes
Easily grown from seed after soaking in boiling water. Needs to be pruned after flowering.

X 0.5

Related Species
(See also *A. genistifolia* p. 44 in this booklet.)
There are several acacias similar to *A. suaveolens* in the drier parts of Tasmania. *A. stricta,* which is very common in dry woodlands in the north, has round flower clusters alone on a single stem.
A. mucronata has long bright-green narrow leaves and flowers in cylindrical spikes. The arrangement of the flowers in cylinders, multiple balls or single balls is important when identifying the wattles.

Acacia genistifolia
Mimosaceae

Spreading Wattle

Distinguishing Features
A shrub with prickly phyllodes (leaf stems which resemble leaves) which are scattered along the stem. A ridge on the branch meets the base of each phyllode.

Description
A very variable shrub, sometimes prostrate, but it may be up to 2m tall with long spreading branches. The spherical flower heads vary in colour from rich yellow to creamy-yellow.

Flowering Period
Spring.

Distribution
Common on dry sites in the east and north.

Locality
Roadsides at the Dazzler Range, south of the Narawntapu National Park. Mt Nelson near Hobart.

Cultivation Notes
Easily grown from seed after soaking in boiling water. May need to be pruned after flowering.

Related Species
(See also *A. suaveolens* p. 43 in this booklet.) There are several acacias similar to this species in the drier parts of Tasmania. Some of these are uncommon like *A.axillaris* which is mainly restricted to river banks. The most common is *A. verticillata* which grows in damp habitats. Its phyllodes are arranged in rings of six around the stem and the flowers are arranged in long spikes, not spherical heads. (See Coastal Plants Identikit).

Pultenaea daphnoides
Fabaceae

Native Daphne

Distinguishing Features
A slender erect shrub with cuneate leaves (leaves which are wedge shaped with the narrow part at the base).

Description
This shrub is often 2m tall. The leaves are dark green above but paler below. Pea-shaped flowers are clustered at the ends of the branches. Each flower is mainly yellow with a red-brown petal at the base.

Flowering Period
Late spring.

Distribution
Common in the more shady parts of woodlands. Also found in a more dense form as a component of coastal heathlands.

Locality
Narawntapu National Park.
Start of the track to Cape Hauy.

Cultivation Notes
Easily grown from seed after soaking in boiling water. Fast growing and apparently short-lived.

X 1.0

Related Species
Twelve species of *Pultenaea* are recorded for Tasmania and many of these are lowland plants. All of the Tasmanian species have flowers which are mainly yellow. *P. pedunculata* is a prostrate shrub with short linear leaves and flowers extending along the branches. It is locally common forming extensive mats. *P. gunnii* has oval leaves with their edges curved back. The small yellow flowers are in dense clusters at the ends of the branches.

Platylobium triangulare
Fabaceae

Distinguishing Features
A weak shrub with triangular leaves. The flowers have a long slender stalk.

Description
This shrub is often found tangled through larger sedges and other vegetation. The leaves have long spines at each of the three angles. The flowers are an attractive red and yellow with old pods often hanging on to the plant long after the seeds have fallen.

Flowering Period
Spring.

X 0.5

Distribution
Mostly restricted to sandy heathlands at a few places on the north and east coasts.

Locality
Wineglass Bay Track, Freycinet National Park. Sandy heathlands near Bridport.

Cultivation Notes
May be grown from seed after soaking in boiling water.

Related Species
The species in the genus *Platylobium* are very distinctive in Tasmania. *P. obtusangulum* is very similar to *P. triangulare*. The only noticeable difference is that the flower stem is surrounded by shiny brown scales. This species extends to dry inland sites, but is also found on coastal heathlands. The third Tasmanian species is *P. formosum* which grows more upright and has a rounded base to its leaves, with just one spine at the end of the leaf. This species is mainly found in dry inland woodlands.

Bossiaea cinerea
Fabaceae

Distinguishing Features

This is a weak shrub which often trails over nearby bushes. The leaves have a rounded base and a sharp point at the tip.

Description

The yellow and brown pea flowers extend along the branches forming a showy display. Each flower occurs singly on a long stalk. Leaf size can be variable and the colour may be grey-green. The pod is more or less rectangular when mature.

Flowering Period

Spring.

Distribution

Common in sandy heathlands and woodlands. Also found in wet forests in the west.

Locality

Peter Murrell Reserve, near Blackmans Bay. Freycinet National Park. St. Helens.

Cultivation Notes

Easily grown from seed after soaking in boiling water. In a well-watered garden it can grow into a dense shrub.

Related Species

There are several species of *Bossiaea* in Tasmania. *B. riparia* is very distinctive with its flattened stems about 1cm in width, and only minute leaves. It is occasional on river banks and dry hillsides. *B. prostrata* is common in dry woodlands. It is an insignificant wiry prostrate shrub and is only seen easily when a few pea flowers are spotted on the ground. *B. cordigera* has prostrate branches, which resemble a fern, and is locally abundant in the north.

Aotus ericoides
Fabaceae

Golden Pea

Distinguishing Features
The crowded narrow leaves are green on top and hairy below, although the under surface is often obscured by the edges of the leaves which are curved under.

Description
Yellow and red pea flowers extend along the branches among the leaves. The pod is distinctly hairy. Usually a dense much-branched shrub up to 1m tall but occasionally prostrate. In the north of the State it may be a very dense shrub with very narrow leaves.

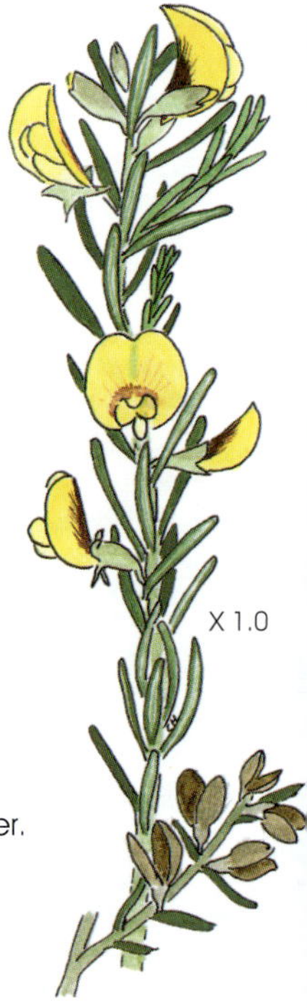

X 1.0

Flowering Period
Late spring.

Distribution
A prominent component of sandy heathlands near the coast around the State. Locally abundant in dry woodlands inland, often on sandy soils. Also found in poor soils on hillsides near the west coast.

Locality
Near the start of the Wineglass Bay Track, Freycinet National Park. Rocky Cape National Park.

Cultivation Notes
Easily grown from seed after soaking in boiling water. May also be grown from cuttings.

Related Species
Many heathland plants are similar to this species when not in flower. For example see *Tetratheca pilosa* in this booklet. No other member of the genus *Aotus* is known in Tasmania, but there are several other members of the pea family similar to this species. The three members of the genus Dillwynia have very fine linear leaves. The common *Dillwynia glaberrima* has flowers which appear to be flattened as they are much wider than high. The orange flowers of *D. sericea* are also slightly flattened.

Diuris sulphurea
Orchidaceae

Tiger Orchid

X 0.5

Distinguishing Features

A yellow orchid with up to six flowers on one stem. There are two brown eye-like spots on top of the bud just before it opens.

Description

A plant which exists during the summer as an underground tuber. A few grass-like leaves emerge in autumn and winter and later a single stem supports the flowers. The six segments which make up the flower are variously shaped.

Flowering Period

Late spring.

Distribution

Common in dry woodlands and sandy heathlands in the east of the State.

Locality

North of The Neck, Bruny Island.
Heathlands south of Waterhouse Point.

Cultivation Notes

Orchid propagation is a specialised field. It may be possible to buy dormant tubers from specialised nurseries.

Related Species

Several species of *Diuris* are common in Tasmania, while a few others are rare. They are all predominantly yellow flowers with brown patches. *D. pardina* (formerly *D. maculata*) flowers earlier than *D. sulphurea*. It is called the leopard orchid because it is typically spotted with brown. *D. orientis* (formerly *D. corymbosa*) is called the wallflower orchid and has brown patches which merge gradually into the yellow. Golden Moths have petals spreading horizontally and have been separated into three species with one, *D. lanceolata* endemic in the far north-west, and the other two more widespread – *D. chryseopsis* in the Midlands and eastern half, and *D. monticola* in highland areas.

Goodenia lanata
Goodeniaceae

Native Primrose

Distinguishing Features
Yellow irregular flowers with winged petals. The leafy stems are usually prostrate, and the flower stems are much longer than the leaves.

Description
This is an inconspicuous plant when not in flower. The dark-green leaves are lobed on the edge and often shiny when fully mature. The flowers are very prominent and quite beautiful when examined closely. They are about 2cm across.

X 0.5

Flowering Period
Late spring and summer.

Distribution
Common in dry heathlands and woodlands.

Locality
Kate Reed Reserve, Launceston.
Ridgeway Park near Hobart.

Cultivation Notes
Easy to propagate from seed. Will often form a more lush plant in a garden environment.

Related Species
Plants in the family Goodeniaceae are very distinctive, having petals which are nearly always winged. Tasmania has several species, most of which have yellow flowers. The only shrubby species is *G. ovata* which is very common in damp forests. Other species are more local, including *G. humilis* which has much smaller flowers and is found in areas subject to flooding, such as the southern shore of Lake St Clair.

Hibbertia riparia

Dilleniaceae

Guinea Flower

Distinguishing Features

Yellow flowers on a plant with narrow linear leaves which are usually shiny. The reverse of the leaf is marked with two narrow indentations from the base to the apex.

Description

One of a large number of heathland shrubs which have narrow linear leaves about 1cm long. The yellow flowers have a group of up to 10 stamens clustered at the centre. In some dry areas the leaves have a grey appearance caused by dense minute hairs.

Flowering Period

Mostly late spring and early summer, but also at other times.

Distribution

Abundant in many places in the eastern part of the State.

Locality

Wineglass Bay Track, Freycinet National Park.
Peter Murrell Reserve near Blackmans Bay.

Cultivation Notes

May be propagated from cuttings. May form an untidy shrub if not pruned carefully.

X 1.0

Related Species

The showy yellow flowers of the genus *Hibbertia* are prominent in woodlands and heathlands. Many have linear leaves, including *H. sericea* which is common around the coasts. In this species the leaves are silky-hairy and usually wider than those of *H. riparia*. *H. prostrata* has fine clustered leaves and as the name suggests it is usually prostrate. Several other species will be seen frequently. A detailed study of the arrangement of the stamens in the centre of the flower is required for accurate identification.

Chrysocephalum apiculatum

(formerly Helichrysum apiculatum)
Asteraceae

Common Everlasting

Distinguishing Features

A small herbaceous plant with white woolly leaves and small clusters of bright yellow flowers.

Description

Plants often form a dense patch of silvery stems and leaves. The leaves are variable up to 5cm long and 1cm in width. The flowering stems are up to 25cm tall. Usually no more than 10 flower heads occur in the terminal cluster.

Flowering Period

Mainly late spring and early summer but also at other times.

Distribution

Dry places in the east of the State, often on cliff edges.

Locality

Tunbridge tip. Hillsides north of Bellerive on Hobart's eastern shore.

Cultivation Notes

Easily grown from seed or cuttings and will grow in well-drained soils.

X 1.0

Related Species

The herbaceous species of the former genus *Helichrysum* have been separated into four genera with a total of 16 species. A number of these, such as *Helichrysum scorpoides* and *Xerochrysum subundulatum* (formerly *Helichrysum acuminatum*), have single flower heads. *Chrysocephalum semipapposum* can be confused with *C. apiculatum* but has greener, narrower leaves and larger flower heads.

Bulbine glauca
Liliaceae

Rock Lily

Distinguishing Features
A tuft of onion-like leaves with a spike of large yellow flowers up to 60cm tall. Each flower may be 3cm across.

Description
This lily is often found growing in a group of similar plants. Usually only a few flowers are out at one time. Careful inspection of the flower will reveal that the stamens are all densely bearded with yellow hairs. The fruit is globular, about 5mm in diameter.

Flowering Period
Late spring.

Distribution
Common in rocky places, especially moss-covered rocky ledges liable to dry out in summer. Also on sand-dunes.

Locality
Tyndall Road, north of Kingston Beach. Waverley Park Nature Reserve, Bellerive.

Cultivation Notes
Easily grown from seed, and regenerates readily from seed in the garden.

X 0.5

Related Species
Bulbine bulbosa is very similar but has a bulb-shaped tuber which is absent in *B. glauca*. The third species in Tasmania is *B. semibarbata*, which is smaller than the other species and has only three of its stamens bearded.

Dianella revoluta

Liliaceae

Blue Flax-lily

Distinguishing Features

A tufted plant with leathery leaves which are folded down the centre and whose edges curve backwards.

Description

Plants of this lily often form large tussocks which produce many flower spikes which are longer than the leaves. Usually only a few flowers are open at the one time. The small blue fruits are pretty but relatively few flowers produce fruits.

Flowering Period

Late spring and early summer. Fruit mainly in the summer.

Distribution

Widespread in dry woodlands and heathlands.

Locality

Southern side of the Cataract Gorge at Launceston. Peter Murrell Reserve, near Blackmans Bay.

Cultivation Notes

Propagation is by division or from the shiny black seed contained in the blue fruits.

Related Species

D. brevicaulis (see Coastal Identikit) is similar but has flower spikes which are shorter than the leaves. It is found mostly in east coastal sites. Four species of *Dianella* are listed for the State. *D. tasmanica* is the most showy, certainly in fruit. It is common in many lowland habitats, but prefers shady moist sites. *D. amoena* occurs at a few dry sites, such as the Tunbridge tip in the Midlands.

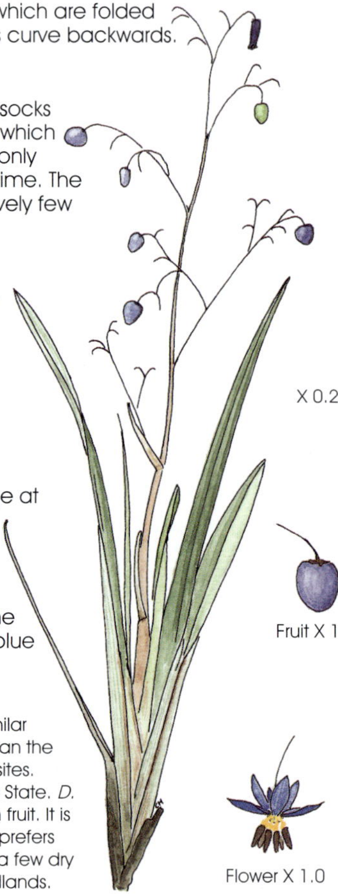

X 0.25

Fruit X 1.0

Flower X 1.0

Wahlenbergia stricta
Campanulaceae

Bluebell

X 0.5

Distinguishing Features
A herb with long flowering branches and blue bell-shaped flowers which are about 3cm in diameter.

Description
A herb which is much-branched at ground level. The leaves are mainly confined to the base of the plant and they are noticeably serrated. The blue flowers are very prominent and have a tube which is about 10mm long and lobes about 15mm long. The outer surface of the flower is paler blue than the inside.

Distribution
Woodlands especially in grassy places.

Locality
Eaglehawk Neck. Epping Forest in the Midlands.

Flowering Period
Summer.

Cultivation Notes
May be propagated from seed.

Related Species
There are about 8 species of *Wahlenbergia* in Tasmania. Most are similar to *W. stricta*, except that several species have much smaller flowers. They are notable for flowering in mid-summer in dry places, adding splashes of blue where there would otherwise be very little colour. The Tasmanian bluebells are related to the *Campanula* or harebell of the northern hemisphere and tropical mountains.

Comesperma volubile
Polygalaceae

Blue Love Creeper

Distinguishing Features
A slender climbing plant with sprays of small blue flowers.

Description
This plant is inconspicuous when not in flower since the green stems have few, if any, leaves. When in flower it is easily seen and very attractive. Fine specimens may climb 2m or more up suitable supporting vegetation. The seeds are contained in small purse-like capsules. Occasionally pink-flowered or white-flowered plants may be found.

Flowering Period
Spring.

Distribution
Common in dry woodlands and sandy heathlands.

Locality
Woodlands near the coast at Southport and Bicheno.

Cultivation Notes
Not commonly cultivated.

Related Species
The other species of *Comesperma* in Tasmania fall into two groups of mostly small erect shrubs. Two have blue flowers and the other two have purple flowers. The most common in sandy heathlands is *C. calymega* which has blue flowers and has only a few yellow-green leaves. Of the purple-flowered species, *C. retusum* is the more common especially on poor soils on the western mountains. *C. retusum* has crowded flat leaves. *C. ericinum* has leaves with the edges curled under, and is found on a few sandy heathlands in the east and north.

Tetratheca pilosa
Tremandraceae

Lilac Bells

Distinguishing Features

A shrub with narrow linear leaves and pink to purple flowers with four petals.

Description

The branches of this small shrub are usually upright, and may be densely clustered. The narrow leaves, with curved-back edges, are similar to the leaves of many other heathland plants. In some locations the leaves are hairy. The flowers are at the base of the leaves and extend along the upper parts of the branches.

X 0.5

Flowering Period

Spring.

Distribution

Woodlands and heathlands throughout the State.

Locality

Near Postmans Pass, Rocky Cape National Park. Freycinet National Park. Tasman Peninsula.

Cultivation Notes

May be propagated from cuttings. Can be short-lived in the garden.

Related Species

The five species of *Tetratheca* in Tasmania all have similar attractive flowers. *T. labillardierei* is a common component of woodlands and heathlands especially on sandy soils and has leaves which are usually finely toothed and wider than the leaves of *T. pilosa*. *T. procumbens* is usually prostrate with small flowers and grows in alpine areas. *T. gunnii* is very rare and inconspicuous. It has been rediscovered only recently near Beaconsfield. Another rare species in Tasmania is *T. ciliata* which may be found at Greens Beach near the Narawntapu National Park.

Stylidium graminifolium

Trigger Plant

Stylidiaceae

Distinguishing Features

A basal rosette of grass-like leaves with one or more spikes of pink or lilac flowers.

Description

An extremely variable perennial herbaceous plant with overwintering basal leaves. The flowers are interesting because one of the petals is modified to form a movable column which contains the male and female parts. Initially it is held away from the flower. It moves quickly towards the centre of the flower when it is irritated by an insect.

Flowering Period

Summer.

Distribution

Very common in many places from sea-level to high mountains. Often in poor soils.

Locality

Freycinet N. P. Peter Murrell Reserve. White-flowered form at Jim Crowe Scrub, Marrawah.

Cultivation Notes

May be propagated from seed. Best grown in pots.

Related Species

The genus *Stylidium* has its stronghold in Western Australia, where there are more than one hundred species with a wide variety of flower form and colour, and a range of growth habit from small plants to a confused tangle of wiry stems. Tasmania has three perennial species. *S. armeria* is a robust plant from damp coastal sites such as Waterfall Bay, and the south coast. *S. dilatatum* is very closely related to *S. graminifolium*, being distinguished by only minor characteristics of the leaves. The other three species are tiny annuals growing for only a few months in places which are wet in spring and dry out in summer.

Indigofera australis
Fabaceae

Native Indigo

Distinguishing Features
Long sprays of showy purple pea flowers at the bases of the leaves.

Description
A shrub usually about 1m tall with few to many slender weak branches. The leaves are divided into up to twenty grey-green leaflets. Sometimes most of the leaves are absent, perhaps having been eaten by pests. The pods are long and slender.

Flowering Period
Spring.

Distribution
Widespread but rarely common, usually in shady places in woodlands.

Locality
Track to Mt Maria, Maria Island National Park.

x 0.5

Cultivation Notes
Easily grown from seed after soaking in boiling water. Needs to be pruned after flowering to avoid becoming straggly.

Related Species
There are about 700 species of *Indigofera* world-wide. About 28 of these species are found in Australia. Two or three species are the source of the deep violet-blue dye known as indigo. In Tasmania, purple pea flowers are unusual. The genus *Hovea* provides the only other common purple pea flowers. *H. linearis* is a small prostrate shrub which is common in dry areas.

Correa reflexa
Rutaceae

Native Fuchsia

Distinguishing Features
A shrub with stiff side branches, often with relatively long sections lacking leaves. The tubular flowers are usually green and tucked under a leaf or two at the end of side branches.

Description
A shrub which is very variable and sometimes prostrate. The leaves are opposite each other, with their upper surface being rough, resembling soft sand-paper. The stems are densely hairy. Red-flowered forms are commonly cultivated but are apparently rare in Tasmania.

Flowering Period
Mainly late autumn to spring.

X 0.75

Distribution
Widespread in dry rocky places and sandy heathlands.

Locality
Chimney Pot Hill, near Hobart. Near George Town for red flowers.

Cultivation Notes
Propagation from cuttings is very easy.

Related Species
Four species of *Correa* occur in the State. Three of these are similar, having green tubular flowers and similarly shaped opposite leaves. *C. lawrenceana* is typically an erect shrub growing in mid-altitude rocky sites, while *C. backhouseana* sprawls over coastal rocks mostly on the west coast and at Rocky Cape National Park. Only *C. alba*, which also grows on the coast, is easy to distinguish as it has white flowers with a short tube.

Dodonaea viscosa ssp. **spatulata** Native Hop

Sapindaceae

Distinguishing Features

Very shiny green leaves on a large spreading shrub. The fruit has three papery wings and is sometimes coloured deep red but may be green.

Description

Male and female flowers occur on separate shrubs which grow to about 4m tall. The leaves become sticky if they are held tightly between two fingers for a few seconds. The small flowers are green.

Flowering Period

Mostly in summer. Mature fruit mostly in the spring.

Distribution

Common in dry rocky situations especially near the coast in the east.

Locality

Cataract Gorge at Launceston.

Cultivation Notes

May be propagated from seed or cuttings. A purple-leaved form, native in New Zealand, is available from nurseries.

Related Species

D. viscosa is found in Asia, America, southern Africa and many islands in the Pacific. There are eight different subspecies recognised in Australia from many places including the inland deserts but only one subspecies extends to Tasmania.

D. filiformis, Tasmania's only other species, is restricted to the State and is local on rocky river banks, for example at the Repulse Dam near Ouse. The leaves are fine filaments but the fruit is hop-like.

Exocarpos cupressiformis

Santalaceae

Native Cherry

Distinguishing Features

A small pine-like tree with fine yellow-green branches and insignificant leaves. Ripe fruits consist of a fleshy red base with a nut sitting on top.

Description

This plant is a parasite on the roots of nearby plants. The branches are usually dense and droop at the ends. The green flowers are minute and clustered on short stems towards the ends of the branches. Few of the flowers mature into fruits.

Flowering Period

Late autumn to early spring.

Distribution

Common, often in dry rocky sites in eucalyptus woodlands in the east.

Locality

Cataract Gorge, Launceston. The start of the track to Cape Hauy.

Cultivation Notes

Difficult to propagate and grow, probably because of its parasitic nature.

Related Species

Tasmania has five species of this parasitic genus. Two prostrate species are restricted to alpine areas. *E. syrticola* is an erect shrub with slight fleshy rounded stems and is restricted to coasts in the south and north of the State. The only widespread species in the lowlands is *E. stricta* which is also an erect shrub with angular branches. The flowers in this species are in tiny rounded clusters, and the fruit stalks are usually white.

Callitris rhomboidea

Oyster Bay Pine

Cupressaceae

Distinguishing Features

A pine-like tree with large clumps of globular woody cones.

Description

This member of the cypress family is often seen as a young tree because it regenerates from fires only by seed. The young branches are green with tiny leaf scales. They sometimes droop very elegantly. Young branches have a faint pine-like smell when crushed.

Fruit
X 0.5

Flowering Period

Pollen-bearing male cones may be seen easily in winter. The female flowers are less obvious.

Distribution

Near the coast from the Tasman Peninsula north to Bicheno. Sometimes in wet sites, other times on dry ridge tops.

Locality

Hazards Beach Track, Freycinet National Park. Track to Cape Hauy.

Cultivation Notes

May easily be grown from seed. Often forms an elegant dense tree in a suburban garden.

Related Species

The one other species of *Callitris* in Tasmania, the South Esk pine, *C. oblonga*, is rare, growing only near a few rivers in the east and at the Cataract Gorge in Launceston. It is a small tree, often with grey-green foliage. The cones are elongated and not round. The trees in the genus *Callitris* are widespread throughout Australia especially in the dry parts.

Branchlet X 1.0

Index to Scientific Names

The species with the symbol * to their left are endemic to Tasmania. They are known to grow naturally only in this State. Former names are in brackets ( ).

Index to Common Names

Australian Plants Society
Tasmania Inc.